STEM
Waterworks

How Do AQUEDUCTS Work?

Greg Roza

PowerKiDS press.

New York

Published in 2017 by The Rosen Publishing Group, Inc.
29 East 21st Street, New York, NY 10010

First Edition

Editor: Greg Roza
Book Design: Mickey Harmon

Photo Credits: Cover, pp. 1–32 (water) elic/Shutterstock.com; cover, pp. 1–32 (pipes) Kovalenko Alexander/Shutterstock.com; cover (image) kavram/Shutterstock.com; p. 5 (inset) Matyas Rehak/Shutterstock.com; pp. 5 (main), 15 (main) Darren J. Bradley/Shutterstock; p. 7 Christian Vinces/Shutterstock.com; p. 9 (main) smartyunknown/Shutterstock.com; p. 9 (inset) https://en.wikipedia.org/wiki/File:LOS_ANGELES_METROPOLITAN_WATER_DISTRICT_AQUEDUCT_ON_HAVASU_LAKE_-_NARA_-_548942.jpg; p. 11 (main) Photography by Steve Kelley aka mudpig/moment/Getty Images; pp. 11 (inset), 23 (main) Jorg Hackemann/Shutterstock.com; p. 15 (inset) ullstein bild/Contributor/Getty Images; p. 17 (main) David Pedre/E+/Getty Images; p. 17 (inset) Chris McGrath/Contributor/Getty Images News/Getty Images; p. 18 Fouad A. Saad/Shutterstock/Shutterstock.com; p. 21 (main) ChameleonsEye/Shutterstock.com; p. 21 (inset) https://en.wikipedia.org/wiki/William_Mulholland#/media/File:William-Mulholland-in-1924.jpg; p. 25 ChinaFotoPress/Contributor/Getty Images; p. 27 JENS SCHLUETER/DDP/Getty Image; p. 29 Thomas Dickson/E+/Getty Images.

Cataloging-in-Publication Data

Names: Roza, Greg.
Title: How do aqueducts work? / Greg Roza.
Description: New York : Powerkids Press, 2016. | Series: STEM waterworks | Includes index.
Identifiers: ISBN 9781499419917 (pbk.) | ISBN 9781499419931 (library bound) | ISBN 9781499419924 (6 pack)
Subjects: LCSH: Water-supply–Juvenile literature. | Aqueducts–Design and construction–Juvenile literature.
Classification: LCC TD348.R69 2016 | DDC 627.13–dc23

Manufactured in the United States of America

CPSIA Compliance Information: Batch #BS16PK: For Further Information contact Rosen Publishing, New York, New York at 1-800-237-9932

Contents

Water Delivery

Aqueducts are pathways constructed to guide water from a source into a populated area. All aqueducts are sloped, which allows gravity to keep water flowing downhill. They often start at a **reservoir**. Reservoirs provide communities with a safe, dependable source of water. They're often located hundreds of miles away from the towns and cities that depend on them. That's where aqueducts come in!

Aqueducts can be very simple. A farmer might dig a simple trench to direct water from a stream to his crops. However, the world's biggest aqueducts were designed and constructed to provide water to millions of people! Scientists and engineers work very hard to make sure large cities get the water their citizens need, even when the closest source of freshwater is hundreds of miles away.

The earliest aqueducts may have just been trenches dug to guide stream water to crops, but people soon found ways to improve them to fulfill the needs of hundreds, then thousands, and finally millions of people.

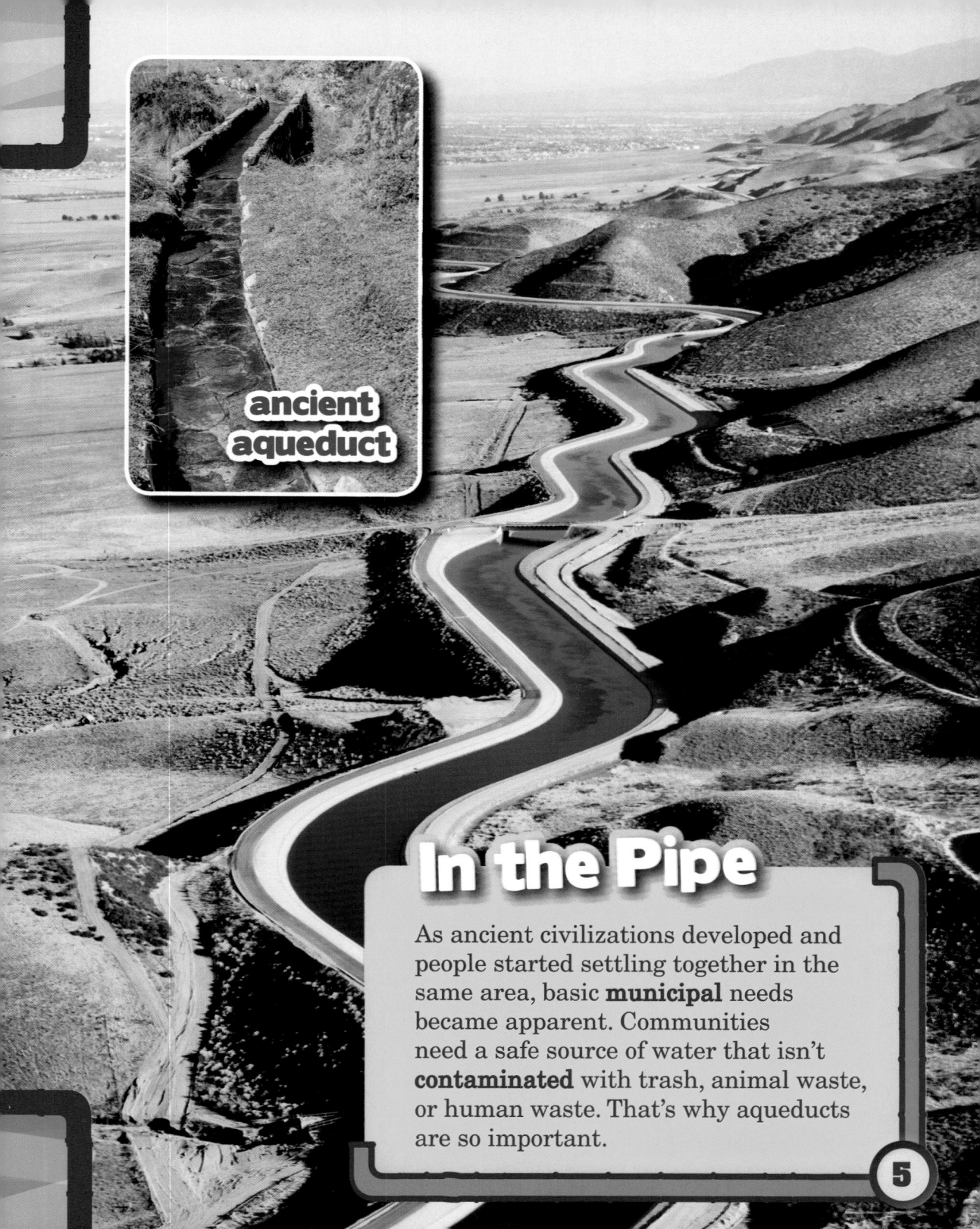

ancient aqueduct

In the Pipe

As ancient civilizations developed and people started settling together in the same area, basic **municipal** needs became apparent. Communities need a safe source of water that isn't **contaminated** with trash, animal waste, or human waste. That's why aqueducts are so important.

Moving Water Throughout History

Archaeologists have found aqueducts in the ruins of many ancient cultures. Aqueducts called *qanāts* collect groundwater in mountainous areas and guide it to lower, **arid** areas for irrigation and drinking. *Qanāts* were probably first used in ancient Persia (Iran) 2,500 to 3,000 years ago. Some *qanāts* are still in use today.

The ancient Romans were aqueduct masters! A total of 11 aqueduct systems were constructed to bring water into the ancient city of Rome. This system served approximately 1 million people. Roman aqueducts are well known for their beautiful arched bridges. These bridges spanned valleys and other low areas while still using gravity to keep the water flowing. Parts of these systems, which were built between 312 BC and AD 226, are still standing.

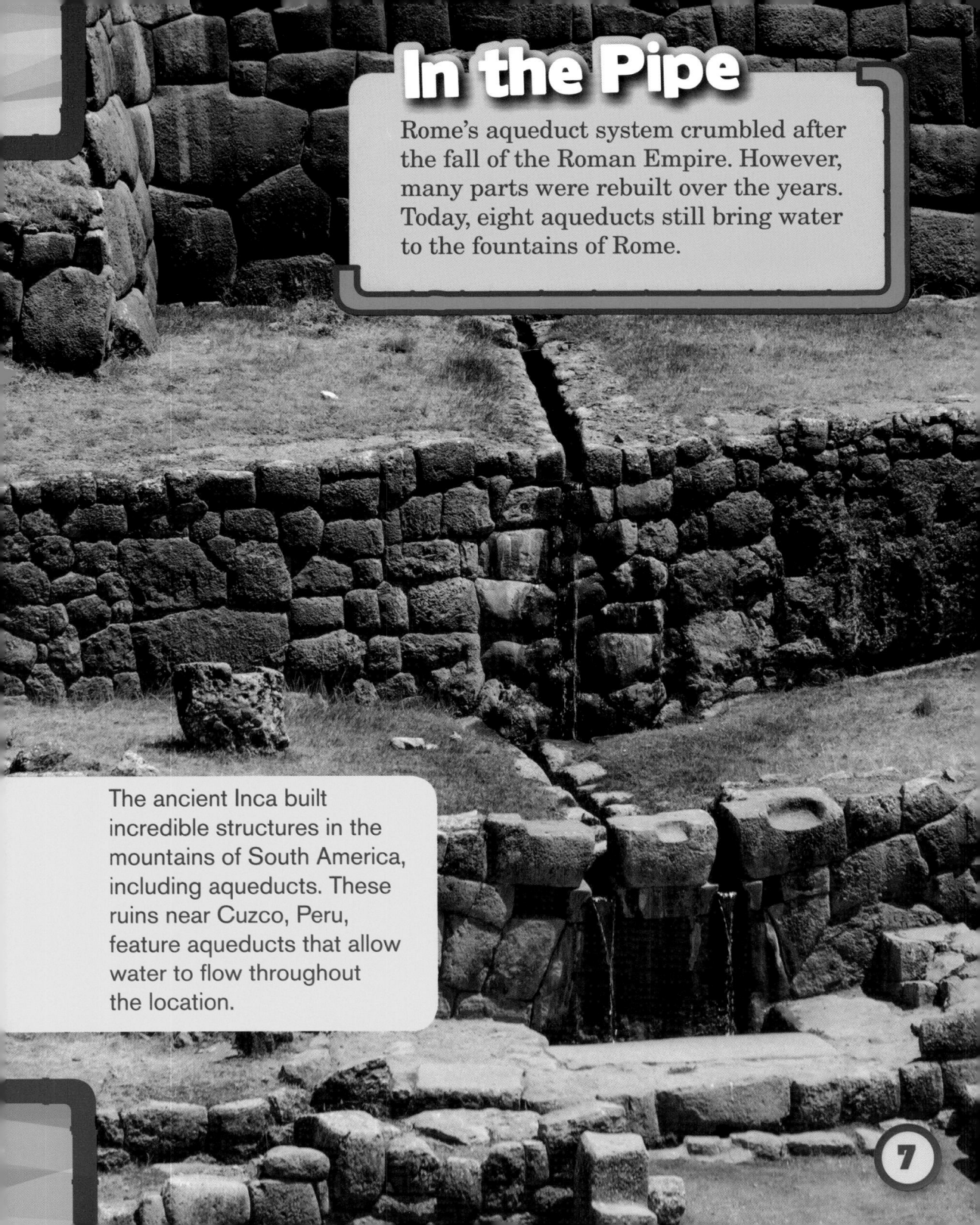

In the Pipe

Rome's aqueduct system crumbled after the fall of the Roman Empire. However, many parts were rebuilt over the years. Today, eight aqueducts still bring water to the fountains of Rome.

The ancient Inca built incredible structures in the mountains of South America, including aqueducts. These ruins near Cuzco, Peru, feature aqueducts that allow water to flow throughout the location.

As cities grew larger, scientists and engineers developed new technology to make better aqueducts. New tools, such as pumps and computers, and new materials, such as steel and concrete, helped make aqueducts more efficient and durable. Some societies used moving water in aqueduct canals to turn waterwheels, which were used to grind grain and do other jobs. Today, giant **turbines** create power for much larger populations.

People build dams to hold water back and create reservoirs. When construction was completed on the Parker Dam in 1938, it created the reservoir Lake Havasu on the Colorado River. A system of pumping stations and aqueducts brings water from Lake Havasu, through desert land, to communities hundreds of miles away, including the city of Los Angeles, California.

Pumps are machines that bring water from a low point to a high point, against the pull of gravity. The pumps shown here take water out of Lake Havasu and allow gravity to move the water through tunnels and open canals until it reaches Southern California.

Parker Dam

9

NYC STEM

Aqueducts require precise calculation and engineering to make sure they work properly and last a long time. It takes many trained professionals to design and construct an aqueduct capable of serving a city's population. Before construction even begins, geologists and **hydrologists** study the environment to make sure construction is safe for plants, animals, and, of course, people. These are just two of the types of scientists needed to construct an aqueduct.

We can learn a lot about STEM (science, technology, engineering, and math) by studying the world's aqueducts, such as the system that serves New York City. In 1776, New York's 22,000 residents got their water from a reservoir. The water was distributed through a system of hollow logs. New York City's aqueduct system today is far more complex.

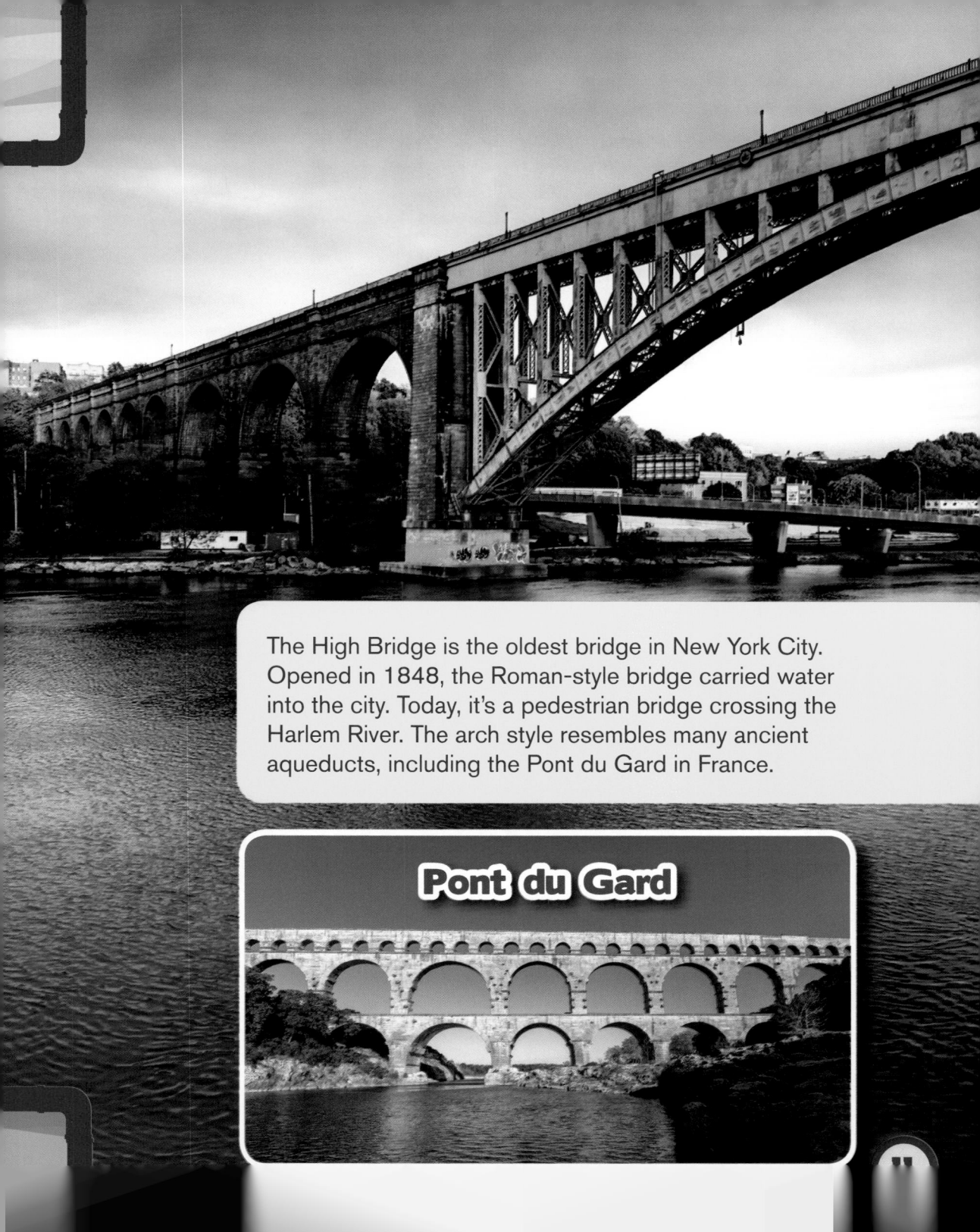

The High Bridge is the oldest bridge in New York City. Opened in 1848, the Roman-style bridge carried water into the city. Today, it's a pedestrian bridge crossing the Harlem River. The arch style resembles many ancient aqueducts, including the Pont du Gard in France.

Pont du Gard

Much of New York City's drinking water is **impounded** in 19 reservoirs and three lakes many miles north of the city. These bodies of water hold about 580 billion gallons (2.2 trillion liters). The three main areas where these reservoirs are located are called **watersheds**.

Since 1842, scientists and engineers have continually added to and improved upon the New York City water system. It now uses a modern marvel of pipes, tunnels, canals, and natural streams to connect the reservoirs to each other and to the citizens of New York. This system makes sure none of the reservoirs run low or overflow. About 95 percent of the water is delivered to New York City by gravity. This saves on energy needed to move the water.

In the Pipe

Scientists and engineers have long known that gravity is a key **component** to any aqueduct. The parts of an aqueduct are designed to keep water always flowing. By using gravity, engineers don't need to use energy to power pumps to move water uphill.

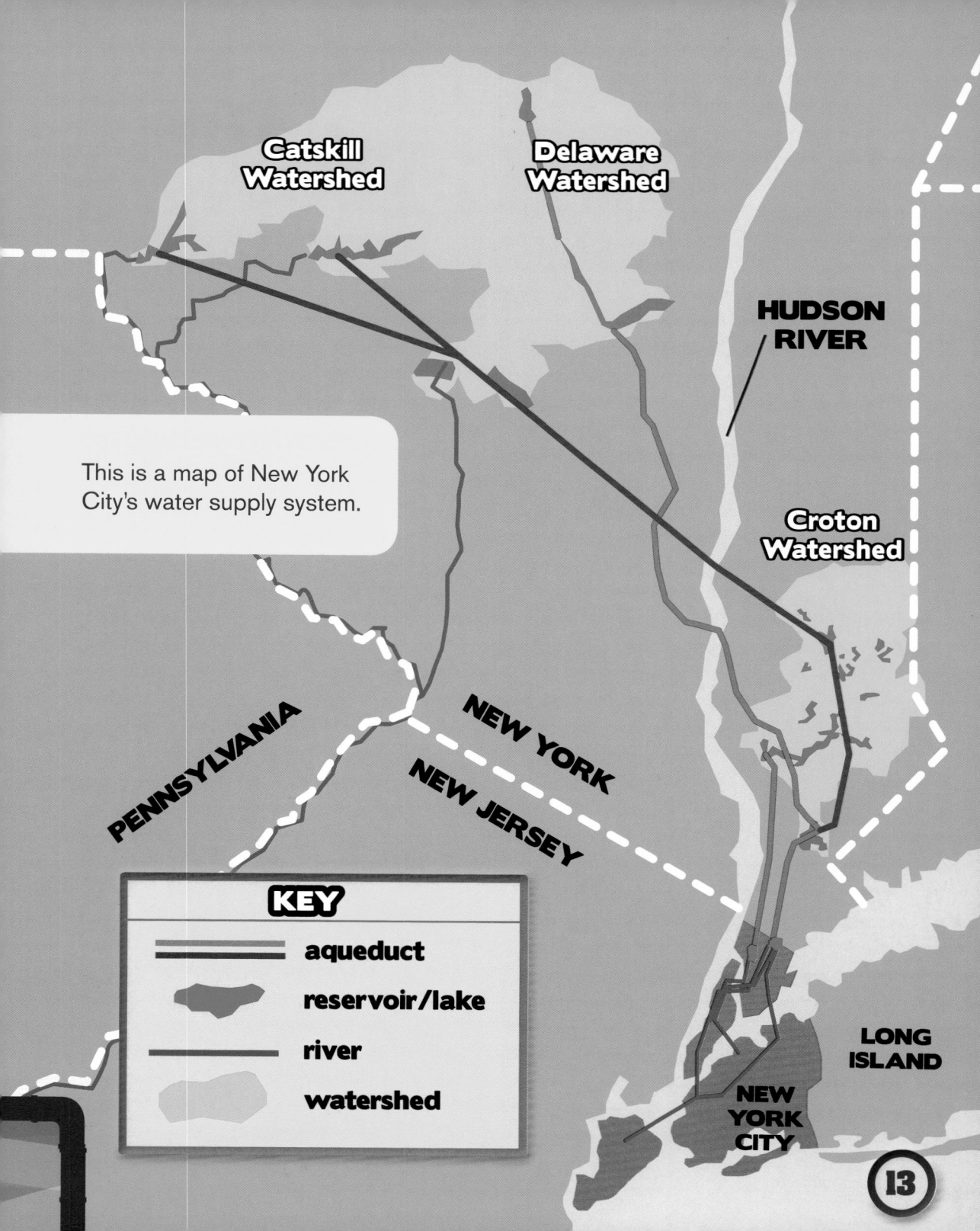

Catskill Watershed

Delaware Watershed

HUDSON RIVER

This is a map of New York City's water supply system.

Croton Watershed

PENNSYLVANIA

NEW YORK

NEW JERSEY

KEY

aqueduct

reservoir/lake

river

watershed

LONG ISLAND

NEW YORK CITY

Technology Talk

Aqueducts both ancient and new use basic technology to move water. Many ancient aqueducts were just open trenches lined with stones. Open canals today are usually lined with concrete. Underground tunnels and pipes are used to move water, too. The ancient Romans mostly used pipes made from terra-cotta and stone, but they also used wood, leather, lead, and bronze. Today, pipes are made from strong metals and plastics.

Although open canals, tunnels, and pipes do most of the work, aqueducts have other important parts. Roman aqueducts made liberal use of arches and bridges to span low areas. Sometimes aqueduct builders include a sharp downhill section. This causes the water to speed up, giving it enough **momentum** to make it to the top of another hill before heading down again. This dip is called a siphon.

open canal

The Los Angeles aqueduct features a siphon running through Jawbone Canyon in the Mojave Desert. Water drops 850 feet (259 m) through a massive steel pipe to the canyon floor. This drop gives the water enough speed to climb back out of the canyon.

Many aqueducts start at a reservoir, and many end at a cistern—a waterproof container for storing water. Some cisterns hold just a few gallons, while others are gigantic. The city of Istanbul, Turkey, has hundreds of ancient underground cisterns. The largest of of these, the Basilica Cistern, is fed by aqueducts that start about 12.4 miles (20 km) away. First built in AD 532, it can hold up to 104,636 cubic yards (80,000 cu m) of water.

Today, aqueduct systems use the best available science and technology. Computer technology is used to design and control aqueducts and water systems. Modern aqueducts may not be as beautiful as ancient Roman aqueducts, but they are far bigger. The California Aqueduct has the longest single aqueduct path in the world, at about 440 miles (708 km).

Japanese cistern

Basilica Cistern

The Basilica Cistern is pretty big. The largest cistern, however, is in Kasukabe, Japan. This modern technological marvel was designed to battle flooding. Massive pumps can push up to 200 tons (181 mt) of water into the Edo River every second.

Engineering Aqueducts

Aqueducts and water systems would not work as well as they do without the hard work of engineers. One of the earliest machines used in water systems is the screw pump. This machine, which is still used today, was probably invented by the ancient Greek engineer Archimedes (c. 287 BC–c. 212 BC). One version of a screw

In the Pipe

During the 1700s and 1800s, newly engineered steam pumps and **pressurized** systems made aqueducts more efficient. Modern aqueducts often have huge pumping stations featuring giant pumps run by electricity.

pump features a pipe with a screw-shaped surface inside it. Other versions feature spiral-shaped pipes. By turning the screw, water is carried up the pipe. This machine was traditionally used for drainage and irrigation.

Ancient Roman engineers constructed beautiful and functional aqueducts. They built bridges with repeating arches. The engineers knew that the arch shape gives their structures strength. They did a great job, because many of these bridges are still standing!

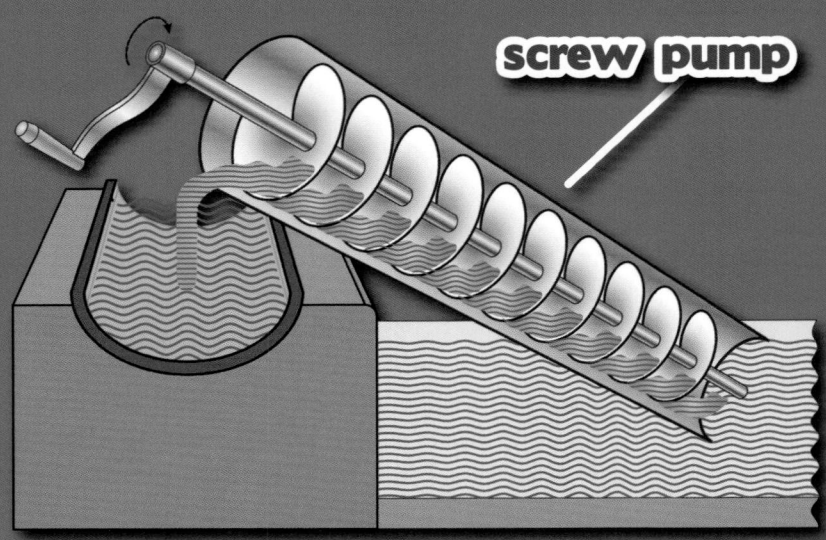

screw pump

This drawing shows how a screw pump works.

Many kinds of engineers are needed to build aqueducts. Water supply engineers are **civil engineers** who help pick the best places for aqueducts and supervise their construction. They make sure the new aqueducts are safe for people and our water reserves.

Chemical engineers are hired to test water reserves for harmful chemicals and to help **purify** our water supplies. They may also help develop new

Civil engineer William Mulholland (1855–1935) became famous for designing and supervising the construction of the Los Angeles Aqueduct. He also designed other water systems, and even gave advice to the builders of the Panama Canal.

materials that stop water from seeping into the ground, such as plastics and ceramics. Mechanical engineers build aqueduct machines, such as pumps and control valves. Electrical engineers are needed to help prepare pumping stations for use and to maintain their electrical systems. Other engineers who might help design, build, and maintain aqueducts include environmental, agricultural, structural, and geological engineers.

In the Pipe

Ancient Roman engineers didn't use chemicals to purify their drinking water. Instead, they built shallow pools called settling basins. There, the water slowed down, which allowed impurities to settle in the basin before the water continued on its path.

Building with Math

Before building a new aqueduct, scientists figure out how much water the system will hold, how much can pass through it at one time, and how steep a siphon needs to be to keep water moving. This is where math comes in!

Calculating gradients, or slopes, is important when designing an aqueduct. If the gradient is too low, the water may not flow fast enough. This could cause it to pool instead of flow. If the gradient is too high, the water may move too quickly. Builders use high-tech surveying tools, as well as a strong background in math, to make sure an aqueduct's gradient is just right. Math is also used to calculate cost and estimate the amount of supplies needed.

In the Pipe

It can be hard to read and write Roman numerals, but imagine how hard it is to calculate with them! Ancient Romans used a tool called a counting board.

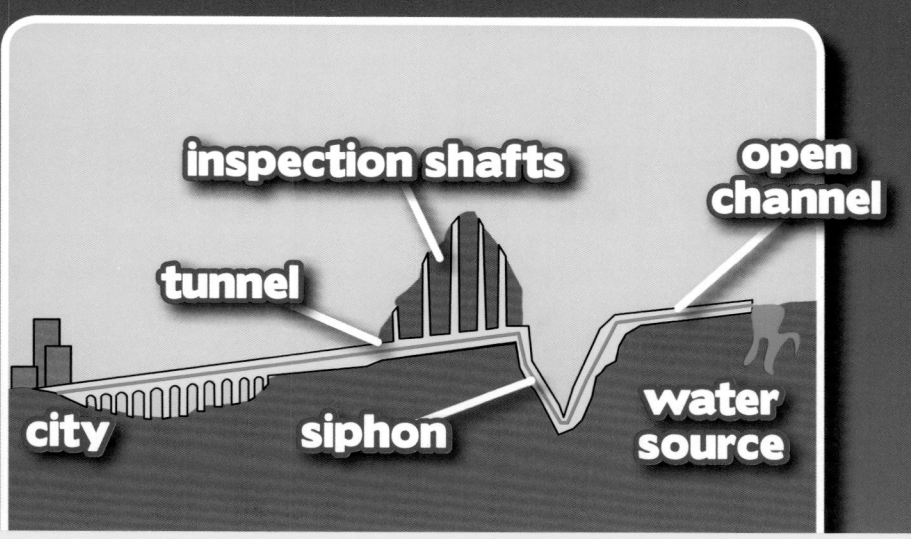

inspection shafts

open channel

tunnel

city

siphon

water source

The ancient Romans understood basic mathematic terms, such as slope and volume. They also understood how to use geometry to build long bridge spans using triangles and arches. When they found a sturdy arch shape, they repeated it many times to make sure their aqueducts would last a long time.

Problems with Aqueducts

Cities in northern China are suffering from water scarcity and polluted water. To ensure people have enough clean water, the Chinese government is building one of the largest aqueduct systems in the world—the South-to-North Diversion Project. This system, when completed, will include China's four major rivers, thousands of miles of aqueducts, and pumping stations capable of filling an Olympic-sized swimming pool in just minutes!

Although China's new aqueduct is sure to solve many problems, some think it will also cause problems. So far, more than 340,000 people have had to move so the aqueduct can be built. Some people think the system will spread water pollution to two of the country's most important rivers—the Han and Yangtze. It might be years before we know the final impact.

The situation in China shows the importance of thorough scientific and environmental research before embarking on such a monumental construction project.

In the Pipe

Different kinds of engineers plan, construct, and maintain reservoirs. Civil engineers build structures such as bridges and dams. Water engineers are civil engineers who focus on structures that deal with water.

Navigable Aqueducts

Not all aqueducts are designed to move just water. Some move people! Navigable aqueducts are designed to carry ships. Sometimes they are built to cross a roadway, and some are built to cross other waterways. These aqueducts are often called water bridges. Engineers sometimes design aqueduct systems using existing streams and rivers. These waterways can also be used for travel and recreation.

At 3,012 feet (918 m) long, the Magdeburg Water Bridge in Germany is the longest navigable aqueduct in the world. It connects the Elbe-Havel Canal and Mittelland Canal and crosses the Elbe River. It was designed to avoid travel problems when the water is low and to make travel quicker for large cargo ships. The route over the water bridge is 7.5 miles (12 km) shorter than the previous route.

The Magdeburg Water Bridge is made of steel and concrete. It's 111.5 feet (34 m) wide and 13.9 feet (4.25 m) deep. This is large enough to accommodate multiple cargo ships at the same time.

In the Pipe

The Netherlands has several notable navigable aqueducts. The Aqueduct Veluwemeer allows small ships to pass over a busy highway! This highway connects the mainland of the Netherlands to Flevoland, which is one of the largest man-made islands in the world.

Engineers have constructed navigable aqueducts with moving parts. Originally built in 1893, the Barton Swing Aqueduct in England allows ships on the Bridgewater Canal to cross over the Manchester Ship Canal. This water bridge can be swung out of the way for ships traveling on the Manchester Ship Canal.

Falkirk, Scotland, features an amazing achievement of modern engineering and technology called the Falkirk Wheel. The Forth and Clyde Canal and the Union Canal were once connected by a series of 11 **locks**. This system was replaced by a much faster rotating boat lift. It's the only one of its kind in the world. Boats enter a section of channel attached to a gigantic wheel. The wheel then turns, either lifting or lowering the boat to the next canal, where it continues on its way.

The Falkirk Wheel has two sections called gondolas, which are where boats sit while waiting for the wheel to finish turning.

Aqueduct Art

The ancient Roman aqueducts were technological marvels, but they were also works of art. The Pont du Gard bridge is part of an aqueduct that's 31 miles (50 km) long. This aqueduct brought water to the Roman city of Nemausus, which is today the French city of Nîmes. The Pont du Gard carried water over the Gardon River. The bridge was considered a technological marvel in its working days. Today, it's considered a beautiful and historic work of art.

You can start experimenting with aqueducts right now. You can use marbles, blocks, paper towel tubes, and other simple objects to experiment with designs. Can you use gravity to get the marble safely to the end of the aqueduct? Use your STEM skills to engineer a fun and functional aqueduct!

Glossary

arid: Receiving little or no rain.

civil engineer: An engineer who designs roads, bridges, dams, and other structures.

component: A part of a larger whole.

contaminated: Polluted.

hydrologist: A scientist who studies the movement and quality of water.

impound: To collect and store water in a reservoir.

lock: A device for raising and lowering ships between stretches of water that are different levels.

momentum: The force that something has when it is moving.

municipal: Having to do with the government of a city or town.

pressurized: Built to maintain an interior pressure greater than the air outside.

purify: To make something pure by removing harmful chemicals.

reservoir: A man-made lake used for storing water.

turbine: A motor operated by the movement of water, steam, or air.

watershed: A low area of land that contains a lake, river, or reservoir into which other rivers flow, creating an important source for freshwater.

Index

A
Archimedes, 18

B
Barton Swing
    Aqueduct, 28
Basilica Cistern, 16, 17

C
California Aqueduct, 16
California, Southern, 9
China, 24, 25
cistern, 16, 17

E
England, 28

F
Falkirk Wheel, 28, 29

G
geologists, 10
Germany, 26
gradients, 22

H
Havasu, Lake, 8, 9
High Bridge, 11
hydrologists, 10

I
Inca, 7

J
Japan, 17
Jawbone Canyon, 15

L
Los Angeles, 8, 15, 20

M
Magdeburg Water Bridge,
    26, 27
Mojave Desert, 15
Mulholland, William, 20

N
Netherlands, 27
New York City, 10, 11,
    12, 13

P
Panama Canal, 20
Parker Dam, 8, 9
Persia, 6
Peru, 7
Pont du Gard, 11, 30

R
reservoir, 4, 8, 10, 12, 13,
    16, 25
Roman, 6, 7, 11, 14, 16,
    19, 21, 22, 23, 30

S
Scotland, 28
screw pump, 18, 19
siphon, 14, 15, 22, 23

T
Turkey, 16

W
watersheds, 12, 13

Websites

Due to the changing nature of Internet links, PowerKids Press has developed an online list of websites related to the subject of this book. This site is updated regularly. Please use this link to access the list:
www.powerkidslinks.com/sww/aque